A COMPOUND OF EXCELSIOR

To John F. Adams
from an admirer
Susan Rice
May, 1996

A Compound of Excelsior

Susan Rice, ASH, BSI

▪ Drawings by THOMAS GENTILLE, BDS
▪ Foreword by THOMAS L. STIX, BSI, ASH
▪ Afterword by EVELYN A. HERZOG, ASH, BSI

Gasogene Press, Ltd.

Dubuque, Iowa

Publication Staff

Jean R. Starr, *Managing Editor*
Renée Meyer, *Wilderness Graphics, Cover*
Gilliland Printing, Inc., *Printing and Binding*

ISBN: 0-938501-14-3

Printed in the United States of America

10 9 8 7 6 5 4 3 2 1

This book is dedicated to the memory of my parents:

Howard Franklin Shout
Ruth Grabowsky Shout

A BEE IN HER BONNET

I have known Susan Rice since we were both younger. She continues to be younger and more beautiful. She is also smart, vivacious, and a hell of a Sherlockian, but when she gets a BEE in her bonnet she can't be stopped. *A Compound of Excelsior* began as many other ventures have. It was an oral presentation to a group of Sherlockians gathered in Kansas City on Labor Day weekend of 1990. David Hammer, the owner of Gasogene Press, was in Susan's audience and persuaded her to expand her talk into a book.

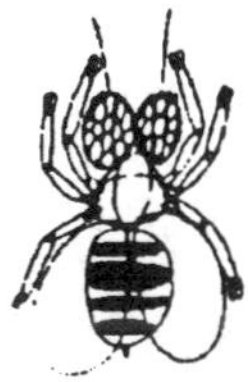

Susan gives us a completely canonical clarification of why the Master retired to take up bee farming. Her logical explanation of the manner in which this came about seems to me to be beyond refutation. Not only has she done her research in the Canon, but she has gone to many outside sources to help the reader understand some of the reasons why bees are so fascinating both to her and Mr. Holmes. A vote of thanks should be given to both the author and the publisher for making this book available to us.

Another BEE in Susan's bonnet is that for the distinguished few who are members of both The Baker Street Irregulars and The Adventuresses of Sherlock Holmes she has proclaimed the acronym ASH should precede the acronym BSI because A comes before B in the alphabet even though BEE comes before A in her affection. It is also true that she is "A BSI" and "AN ASH." Obviously I believe BSI should precede ASH. It did by some thirty-five years.

Thomas L. Stix, Jr., BSI, ASH
Wiggins

Author's Note: Mr. Stix sent me this foreword scant days before he learned that he was one of the distinguished few to hold both investitures, and that some questions have more than one answer.

ACKNOWLEDGEMENTS

There are so many joys in being part of a community of Sherlockians that we sometimes forget that it is also a community of scholars. It is a generous community as well, and every Sherlockian of whom I asked a question sought out and brought forth answers with willing speed. Life is a lot easier when you have friends at the top. For advice in matters Holmesian and honeybee, I'd like to thank Mickey Fromkin, always first editor; Eileen Hartsoe, M.D. and Bob Katz, M.D. who advised me on medical matters; Evelyn Herzog on whom I called for expertise in Holmes and in Latin; Tom Stix for his thoughtful comments and suggestions; and Joanne Zahorsky and Roger A. Morse of Cornell University who turned their prismatic eyes on my beeish thinking. Dr. Morse's musings on what aspects of bee culture Holmes explored have

tempted me to further research already. For gifts of books that eased and broadened my research I give thanks to Mickey Fromkin, Evelyn Herzog, Sheila Hickey, Sarah Montague, Dore Nash, Warren Randall, Mary Ellen Rich, Tom and Ruthann Stetak, Thom and Maryellen Utecht, and Richard Wein (whose generous gift of *the only copy he found* of Reverend Wood's cornerstone work proved he's a friend before he's a collector!). And to participate in any venture with a friend and artist of the stature of Thomas Gentille is a privilege, indeed. Much of the book's design was conceived by him.

I'd also like to offer sincere thanks to David L. Hammer and Jean R. Starr, publisher and editor respectively of Gasogene Press. Their willingness to cooperate in creating the unconventional book I saw in my mind's eye made me the luckiest of neophyte authors. I thank them for their patience with my eccentric vision.

I'd also like to thank John Bennett Shaw. He has given no direct assistance in this project, but without him there would be no book.

Although my acquaintance with science has been nodding and not always cordial, I have found the study of bees to be absolutely riveting. It is still possible, even at this remove, to glimpse in the science of apiculture that which fascinated Mr. Holmes. Sauntering down the byways of bee culture is one of the most pleasant routes I've found to feeling more closely akin to the Master.

I hope to tempt you to seek out the books of the great Victorian bee masters, and to that end I include an annotated bibliography. For those I cannot persuade to further reading, I have tried to find the words to make clear to you the excitement of the mysteries of the honeybee. In my attempt to communicate this excitement, I have elected to eschew footnotes because I fear their power to distract. Every book I cite is in the bibliography, and I've tried to attribute quotations clearly within the text. An additional Sherlockian bibliography follows the appendix.

"Babbles the bee in a stolid ear"

Emily Dickinson
"Safe in Their Alabaster Chambers"

It's exhilarating to consider the reactions of those early readers of Sherlock Holmes, learning of his exploits while the ink was still fresh on the pages of *Strand Magazine.* Somehow, we feel, the joy would have been purer for them, but so, let us remember, would the terror. Assuming we first met Holmes in one of the omnibus editions, we read "The Final Problem" in the approximate center of a very large book. We must have thought: if Watson's best and wisest man had been tipped into the Falls, then what the devil is contained in the rest of the volume? But imagine the reader of 1893 learning of the Master's death in the Reichenbach without the reassurance of the unread pages ahead. What a traumatic moment for a devotee! We've heard that bereaved readers wore crepe armbands and wrote furious letters to Dr. Arthur Conan Doyle, but stop and consider for a moment the feeling of that contemporaneous reader—how his or her heart must have fallen, far deeper than the Falls near Meiringen. In contrast, it's fun to think of the aficionado just after the turn of the century reading as the last words of another month's installment in *Strand Magazine,* "Mr Holmes, they were the footprints of a gigantic hound!"—and then having to close the magazine and wait thirty endless days to learn more. And that period of excruciating impatience assumes that the *Strand* of yore was far more timely than most Holmesian publications today.

Little attention has been given, however, to the reaction of the reader as his or her eye fell on the astonishing news contained in the preface to *His Last Bow,* first published in 1917:

> The friends of Mr. Sherlock Holmes will be glad to learn that he is still alive and well, though somewhat crippled by occasional attacks of rheumatism. He has, for many years, lived in a small farm upon the downs five miles from Eastbourne, where his time is divided between philosophy and agriculture.

Note how this vital information comes to us, not as part of the action of the story, but as a casual aside to readers, a throwaway, if you will, placed at the beginning of the story, at the front of the whole volume even, so its impact is cushioned by the narrative ahead. Just as we learn there will be no more cases, we can see a new adventure starting just below. The reader's eye is drawn irresistibly to the fast-approaching conundrum at Wisteria Lodge. But perhaps when the end of the tale has been played out, and the honorable Aloysius has surrendered his life for his country, perhaps then we have time to stop and reflect on the remarkable piece of information we were tendered before we were grabbed by the story.

An examination of the higher criticism tackling the question of Holmes's retirement yields surprisingly little of substance. A careful search has uncovered only a handful of papers attempting to deal with the most basic puzzle: why, at the height of his career and in the prime of life, did Sherlock Holmes retire? Lots of loquacious Sherlockians with wide-ranging opinions have failed to comment on a period of our Master's life that has lasted 88 years to date. Clearly, the years that follow 1903 are uncomfortable ones for most of us. We must face our discomfort and turn a steady gaze on Holmes's decision to retire from his career as a consulting detective. So let us return to the words of the preface, and the early years of what has been labelled Holmes's retirement.

Though it may appear didactic, we can separate the startling news vouchsafed us into four distinct segments. We will consider each portion separately: first, the statement that Holmes has left London and is living on the Downs; second, that he has retired from his career as consulting detective; third, that his time is engaged by philosophy and agriculture; and fourth, that he suffers from rheumatism.

We are first told that Holmes has spurned life in our beloved cesspool to reside in an isolated cottage on the Sussex coast. This seems hard to believe of the man who found such stimulation in the endless variety and possibility of the great city. This is the man of whom Watson remarked in "The Adventure of the Cardboard Box":

> . . . neither the country nor the sea presented the slightest attraction to him. He loved to lie in the very centre of five millions of people, with his filaments stretching out and running through them, responsive to every little rumour or suspicion of unsolved crime.

We recall his enjoyment as he gazed at the throngs in the street, rising to the challenge of making instant observations and deductions as they passed. What secret was hidden by that coat collar? What was revealed by that trouser cuff or the shape of that hand? He seemed to revel in the mountains of data produced by the busy city, so it seems incredible that he would seek to isolate himself, cut himself off from all that stimulated his brain and made his life worthwhile. His pleasure with London, his delight in life within the bustling metropolis, was clearly related to his unusual career.

And let us not forget that he must also sacrifice the opera houses and concert halls of the capital. Wilhelmine Norman-Neruda and the brothers de Reszke were not in the habit of performing at the Eastbourne College for Boys. It's difficult to picture Holmes waving his long thin fingers in accompaniment of a sentimental ballad quavered by a local belle, or leaning back with closed eyes to concentrate on a martial tune tootled by a provincial band. When Holmes left London he sacrificed a great deal that made his life pleasurable.

He sacrificed Watson, too, since Holmes admits in "The Adventure of the Lion's Mane" that the good doctor had passed almost beyond his ken. It's true that Watson had selfishly left Holmes for a wife, name and number unknown, but the Master's choice of a new residence decreased even the chance of an occasional visit. Holmes's apparent willingness to trade London for a small cottage five miles from the nearest village certainly seems unaccountable. It must have been something quite remarkable that caused him to depart from all the conveniences and pleasures of London.

Of course, Holmes was a perceptive and observant detective even in the country—we see him frequently functioning with full power on country cases. But he was never distracted from the case at hand by mere bucolic beauty. We remember him squelching Watson's enthusiasm at the freshness of the countryside in "The Adventure of the Copper Beeches":

> . . . They [country houses] always fill me with a certain horror. It is my belief, Watson, founded upon my experience, that the lowest and vilest alleys in London do not present a more dreadful record of sin than does the smiling and beautiful countryside.

Granted, part of this outlook is the Master's remarkable ability to focus completely upon the object of his attention, in this case, crime. But while it seems to most of us that the Sherlock Holmes we know is rooted in London, perhaps we must admit that his departure from London is much less an aberration than his retirement from detection, and indeed, one may relate to the other. Let us proceed, then, to an examination of the second point, his actual retirement, in hope that we may discover an answer to the first.

The preface tells us that Holmes no longer practices the career he invented for himself. He has rendered useless the skills and knowledge he had painstakingly acquired. Of what use now are his brain attic and library lumber-room stocked with such care and precision? It required decades to amass, categorize, and collate the enormous volume of information Holmes drew on in his career as a consulting detective. It is astonishing that he seems so willing to turn his back on the criminals he worked so hard to defeat, to desert the science of detection after he virtually invented it.

Of course, we must acknowledge that Holmes said that his occupation perished with the pernicious Professor Moriarty, but that was years before; surely he has found some mettle fit for his mental sword since then. To that point, we are all aware that Holmes grumbled more than once about the trivial and commonplace nature of the criminal cases with which he was confronted, but it deserves mention that the first complaints of that sort came to the surface as early as *A Study in Scarlet* and *The Sign of Four.* In other words, by 1881 at the latest, the Master was already finding his cases weary, stale, flat, and only occasionally profitable. His retirement is dated in 1903, so for at least 22 years Holmes made the same complaint. While it's difficult to disparage any statement from the Master, it must be admitted that the longevity of this lamentation moves it away from a specific and concrete statement of discontent to something that from anyone else would resemble a whine. That doesn't make it untrue—it's clear that Holmes did not suffer foolishness gladly—but it's also clear that it was not a complaint on which he felt compelled to act.

Discounting this rather continuous refrain about the banality of his cases, can we find other evidence on the likelihood that Sherlock Holmes would be moved to retire? Rather the reverse! Recall some of the things he said over his years of active practice. In *The Sign of Four* he tells us, "My mind rebels at stagnation." And does this sound like a man anticipating retirement: "I never remember feeling tired by work, though idleness exhausts me completely." Indeed, three times he describes what he calls "the insufferable fatigues of idleness" in "The Adventure of Wisteria Lodge." Engraved in all our memories is his famous simile, also from "Wisteria Lodge": "My mind is like a racing engine, tearing itself to pieces because it is not connected up with the work for which it was built." Are these the words of a man eager to disconnect himself from his profession? We remember, too, that in "The Red-Headed League" he quotes with approbation Flaubert's statement that "The man is nothing, the work everything."

He's certainly made it clear that retirement as we think of it is not for him. Idleness would not bring him pleasure; rather, our picture of Holmes is of a greyhound straining in the slip. But if we accept that he would never choose to stop working, then the only meaning we can assign to the word **retirement** is the end of his work as a consulting detective. Perhaps he started a new career.

A strong case can be made that Holmes chose to become a consulting detective because it gave him a greater release from boredom than anything else. Life posed problems for him and he solved them, and thus he staved off the worst of his demons: tedium and ennui. It seemed to him that the whole art of detection offered more challenge than anything else that confronted his remarkable brain. So he wouldn't retire unless he found something better: something that required even greater concentration, a grander escape from boredom. And he was far too intelligent to return to cocaine, however mild the solution.

Whatever it was that Holmes turned to would have to be truly fascinating to absorb that phenomenal brain. Unless it were thoroughly important to him, it could not justify his departure from London and his desertion of detection. And (here comes the third point) our Canon tells us what he turned to. It's just hard for us to accept at first reading, for while we are still staggered by the news that our hero has departed from the active practice of detection at the unusually early age of forty-nine or fifty, it begins to enter our conscious minds that he has given up detection for . . . agriculture? This man of unequalled intellect, the keenest mind of his, and probably our, generation, has retired to become . . . a farmer? And not merely a farmer, for our reeling brains recall another fragment, a telling statement from the first paragraph of "The Adventure of the Second Stain":

". . . since he has definitely retired from London and betaken himself to study and bee-farming on the Sussex Downs . . ." If we are to believe our eyes, Sherlock Holmes left the practice of crime detection to become a bee-farmer. And, as far as we know, he's keeping bees still.

At first, perhaps, we believe it must be a joke—or a code! Surely he isn't really keeping bees . . . beekeeping must be a code phrase meaning something far more intellectual and adventurous than entomology. Studying insects? Our Master? It can't be true. It seems an entirely unworthy pursuit for the man under whose banner we proudly march. Our hero, the beekeeper?

We have focussed on the announcement printed as a preface to *His Last Bow* because it is the baldest and most absolute, but it is actually the last of eight statements about the Master's retirement. Holmes first mentioned the possibility of retirement in 1891 when he was a mere pup of thirty-seven. This is in "The Final Problem," and he speaks lightly, almost jokingly, anticipating a leisurely exploration of chemical researches. Clearly, he has not yet decided on beekeeping at this time. Later in the same story he makes a more telling remark,

> Of late I have been tempted to look into the problems furnished by nature rather than those more superficial ones for which our artificial state of society is responsible.

He was facing death at that time and he knew it, and suddenly the mysteries with which he had been occupying his mind were swept away in his coming confrontation with the last and greatest mystery. In "The Adventure of the Abbey Grange," usually dated in 1897, he talks about his intention to write a textbook to be called *The Whole Art of Detection* when he retires from active practice. "The Adventure of the Creeping Man" is the last case before his retirement in 1903. In it he chides himself for being less astute than he should have been by saying, "it's time for that little farm of my dreams." At last, the fourth time he speaks of retirement, Holmes mentions bee-farming for the first time. We alluded to this quotation earlier; it occurs in "The Adventure of the Second Stain." The next comments come from "The Adventure of the Lion's Mane" and are particularly significant because that case shows us an already-retired Holmes in the cottage that

is the locale of his new life. He surprises us in the first paragraph by writing,

> It occurred after my withdrawal to my little Sussex home, when I had given myself up entirely to that soothing life of Nature for which I had so often yearned during the long years spent amid the gloom of London.

Shortly thereafter he states explicitly, "My house is lonely. I, my old housekeeper, and my bees have the estate all to ourselves." And finally, in the Master's last recorded case, "His Last Bow," Holmes responds to Von Bork's threats to his life by saying,

> The old, sweet song. How often have I heard it in days gone by. It was a favorite ditty of the late lamented Professor Moriarty. Colonel Sebastian Moran has also been known to warble it. And yet I live and keep bees upon the Sussex Downs.

And that brings us to the information in the preface to *His Last Bow,* which alludes to philosophy and agriculture.

Now, while our minds are still buzzing with those eight citations, let us make some observations about them. Two of the earliest spoken, it is generally believed in 1891 and 1897, make no mention of beekeeping and, indeed, project entirely different activities. At the time he speaks those words, Holmes has not yet centered on bees, but in 1891, when he knows he is facing the possible end of his life at the hands of Professor Moriarty, he clearly recognizes that nature poses the greatest puzzle for him. He contemplates turning away from the problems of society toward the deeper ones of nature, though it seems he has not yet decided what precincts of nature to explore. Later, he muses on "the little farm of my dreams," which is, of course, perfectly congruous with bee-farming. From then on he cites beekeeping as his second career, up to the final reference when Holmes is described as spending his time in philosophy and agriculture, again a reference entirely in keeping with bee-farming, which combines them both. Now, Watson as an author and an historian is not given to feats of allusion or sleight of pen. Despite the strenuous attempts of some of his latter-day critics to deny it, he can always be depended upon to mean what he says. Unless we are prepared to cite John Watson as a

liar, repeatedly a liar, then we must accept that the world's first consulting detective is now the world's most intelligent beekeeper.

There is one other point of proof, in this case an absolutely concrete point. In "His Last Bow," masquerading as a notebook of naval signals, is a slim blue volume lettered in gold. Its title, you may recall, is *A Practical Handbook of Bee Culture with Some Observations Upon the Segregation of the Queen,* a fine if somewhat unwieldy title. The Master describes it as "the fruit of pensive nights and laborious days when I watched the little working gangs as once I watched the criminal world of London." Unless we accept the existence of this book, we are in danger of sacrificing the very basis of our scholarship. If we deny the reality of Holmes, the beekeeper and author of this practical handbook, we must label Watson, if he is indeed the writer of this tale, an unmitigated liar, for he invented, quite unnecessarily, a physical object. There can be no claims of mistakes, of hearing Holmes imperfectly or interpreting him wrongly, or inscribing his words in unreadable handwriting. If Holmes is not a beekeeper then Watson is a scoundrel!

Only honey can render this difficult pill easy to swallow. Sherlock Holmes retired from the practice of detection at the age of fifty, give or take a month or two, and became a beekeeper. The internal evidence is convincing, but the truth still seems disappointing. When Holmes departs from the career he invented, we want him to move on to something greater. It seems important for us to believe that beekeeping was the tip of the honeycomb—the outward and obvious manifestation of something bigger, much bigger—a system—a solution to a great problem—something that makes a difference. And our belief need not remain unsatisfied if we can open our minds and see beekeeping as he did.

Inter omnia insecta principatus apibus, et jure praecipua admiratio.

Pliny, Lib II c3
(Among all insects the bee is sovereign, and is due admiration.)

The hive of the bee is a perfect social instrument. Quite simply, honeybees cannot live outside of it. Ostracism virtually always means death—what echoes of ancient Greece. The bees exist together in total and mysterious harmony, each with a clear understanding of her or his place in the world, responding to changing circumstances with what seems to be a single will. Indeed, Maurice Maeterlinck (yes, the same man who gave us *The Bluebird*) in his well-known book about bee culture, called it "the will of the hive." George DeClyver Curtis, on the other hand, thought of it as "the great bee Law of Unity, one hive, one queen, one purpose."

It's truly a wonder: 60,000 or more bees labor together under one roof for the common good. No one shirks, and when work is needed in a particular area, that work seems to call together the workers required to perform it. Yet all this is accomplished with no apparent direction; there are no bosses, no committees, no foremen. In the words of C. N. Buzzard,

> Bees in quantity, a hive of bees, seem to be governed by what I have called the spirit of the hive, which appears to be a power invested with considerable acumen.

Three classes of bees are represented in every functioning hive. Hundreds of drones (male bees) exist for a single sexual act that only a small fraction will experience; the single queen in each hive has only one function: to lay eggs; everything else in the hardworking colony is accomplished by the workers, females with atrophied or incomplete sexual equipment. One measure of the strenuous nature of a worker bee's life is the variability of her life expectancy. A worker born in the summer when the hive is busiest lives approximately four weeks; a worker born in the quiet autumn can live five or six months.

The male bee or drone is a clumsy, large-eyed bee of very little brain. Karl von Frisch, who won a Nobel Prize for his apian explorations, enjoys a chuckle when he says,

> The brain of the drone is smaller than that of both worker and queen—we are not left in any doubt as to the intellectual inferiority of the male in this case.

Since they are born of unfertilized eggs, drones have grandfathers, but no fathers; they are exact genetic copies of the queen, except that they are male. They are much larger than workers, but their bodies are not equipped for any of the work of the hive. Drones cannot forage or clean, and since they have no sting, they cannot defend the hive. Apropos of that, Murray Hoyt remarked,

> In the winged insect world, males don't amount to much. Nature must have felt that giving anything that unimportant a method of defense was wasteful.

German poet Wilhelm Busch called them, "lazy, stupid, fat, and greedy," which brings to mind the fact that Bertie Wooster belonged to the Drones Club. All they can really do is hover on the corner, watching for a queen to fly by.

The drone has one moment of usefulness only, after which he dies. It is his task to fertilize the queen—several called this act "fecundation," and a German bee writer named it "the queen's wedding excursion." This happens just once, and always on a fine, warm day. The young virgin queen flies high up into the air for the first, and possibly, only time in her life. She spirals through the blue, mating with any drone she encounters. The drone arches backward at the moment of climax, leaving his genitalia inside the queen, and falls back to earth to die. The vigorous queen may choose to mate again with another high-flying drone. She stores all the sperm received in this single vaulting flight, and then returns to the hive, which in all likelihood she will never leave again. She will spend the rest of her life, frequently three or four years, laying eggs from this single flight of fertilization.

The fact that mating can take place only in the free and open air prevents the inbreeding which would result were the queen to mate invariably with drones of her own hive. The chances are that the drones she encounters high in the sky will be from another hive, regardless of whether the queen in question is a wild or a kept bee. Attempts have been made to mate queens with selected drones in tall glass houses or huge wire cages, but they have always failed. It seems that unrestricted flight is an essential part of the function. Artificial insemination, on the other hand, calls for no consent and was performed as early as 1927 by a scientist at Alfred University bearing the distinguished name of Dr. Watson. Equipment for this tiny surgical procedure was available to the commercial world by the late forties. Mr. Buzzard speaks of the queen when she returns from her nuptial flight:

> But thereafter the royal prerogatives hardly come up to expectations, as the queen, made a widow on the first day of her glorious honeymoon, must continue amidst the worship and attentions of her faithful subjects, an incessant process of child-bearing without rest or interlude for the remainder of her days, literally the sole mother of her nation.

In prime breeding season a healthy queen will lay 2000 to 3000 eggs a day, each in a perfect hexagonal cell in the honeycomb. Dr. von Frisch computed that she lays, on the average, one egg per minute by day and night. Actually, she takes periods of rest and lays correspondingly faster in between. She can lay twice her own weight in eggs every day. One can see how fast the eggs must grow inside their mother, and it becomes clear why the queen is not free for any other occupation. She lays outward in a spiral, virtually never missing a cell or laying in one twice, and keeps the eggs clustered for efficient temperature control. On the status of the queen within the hive, William Longgood makes the following observations:

> Her position in the hive is ambiguous at best. Despite her own dependency and perilous position, the colony is dependent upon her for its existence. She is indispensible and yet subject to being killed in panic or by whim, or replaced in what amounts to star chamber proceedings against her. Any figurative crown she wears must rest uneasily on her head. Her true role is less that of a queen than mother of the hive, a title often accorded her. And yet, this is something of a mockery because of her lack of maternal instincts or the ability to care for her young. She is, more than anything else, an automated egg-laying machine, a

derogatory but descriptive name
she must bear.

The queen can actually decide the gender of her offspring by allowing each egg to make contact with the sperm she stores or not. She has the power to lay unfertilized eggs which mature within days to become drones, or to lay fertilized eggs which will become worker bees. If the queen confronts another queen, either a hatchling or a reigning queen, one of two actions will ensue. Either they will fight to the death, or the present queen, sensing the presence of the new queen, will gather most of the hive and swarm, departing to seek new quarters. Except in rare and short-lived circumstances, there can be only one queen bee in a hive. Worker bees are the only other class within the hive, and are exceedingly well-named. The drone exists only to fertilize the queen, and after her single hour in the sun the queen does nothing but lay eggs. Everything else is done by the worker bees. They gather the nectar in their honey-stomachs and the pollen in little sacs on their rear legs, deposit both in cells in the hive which they've also constructed, and cap each cell at the proper time to maintain these substances. The nectar, after its transport in the bee's honey-stomach and a certain

measure of evaporation, miraculously reveals itself as honey, a wonder on its own account. The pollen is used to produce bee-bread, the protein-rich substance with which the larvae are fed. The workers build the honeycomb from wax produced by their bodies, weatherize the hive, maintain its temperature and cleanliness, and defend it against foe. The drones have no sting at all; the queen possesses a sting, but, as Moses Quinby words it, "will not use it upon anything below royalty." The workers alone have the means to defend the hive against its many predators. They feed the eggs and maintain the queen. They care for her in every way as she lays the eggs that assure the future of the hive: they feed her, they clean her, they stroke her and remove her waste. Communication is constant between the workers and the queen, who is also their mother. They are continuously exchanging information about the well-being of the hive, and if the queen is not at her peak and the workers believe she is fading, they institute supersedure, the process by which they raise a new queen to replace the old one. They do this by elongating cells that hold newly-laid fertilized eggs in order to encompass the larger queen larva, and then feed that larva an exclusive diet of rich royal jelly, another

substance the workers extrude from their bodies. They feed the queen egg continually and care for it with great attentiveness. Eventually, a new queen will emerge, created by the royal jelly. Mr. Buzzard asks us to give careful consideration to the remarkable process because,

> Worker bees are the only creatures known which can create a superior being of their own genus, anatomically different from themselves, by varying the diet of a young larva.

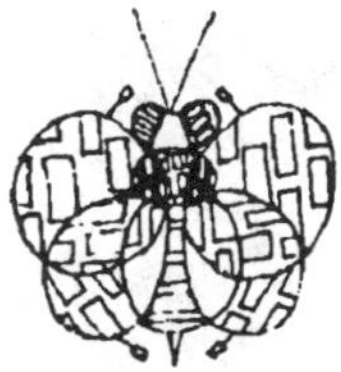

Nature provides no other example of this transformation: a creature changed after birth, after the genetic properties are established, by diet alone. Consider for a moment the physiology of a worker bee. Approximately the size of your thumbnail, that little body performs the following tasks:

- —produces wax to build the cells
- —transforms plant resin into propolis, a substance used to seal the hive
- —manufactures venom to protect the hive

—forms glue on its footpads permitting it to walk on the smoothest surfaces
—gathers and transforms pollen into protein-rich bee-bread to nourish the young
—produces royal jelly and, of course,
—transforms nectar into golden honey

If the shade of William Shakespeare will pardon the pun, it is absolutely astonishing how so many diverse substances are taken into this tiny body and . . . suffer a bee change/Into something rich and strange. Clearly, we are dealing with an extremely complex organism.

The remarkable scope of her duties keeps the bee as busy as the old adage claims. She works so quickly that continuous observation is impossible and even the most attentive apiarist is forced to watch each task in stages. Mr. Hoyt describes a worker approaching a pollen-rich flower with each part of her tiny body operating with perfect efficiency. Everything is in motion: antenna, antenna cleaners, three pairs of legs, all gathering pollen, moistening it, and forming compact loaves.

He observes,

> . . . all these things are going on at once. Things are being combed and transferred, and chewed, and moistened, and packed almost like sleight of hand. Each set of legs is moving independently of the others. Add the fact that the performance can take place while the bee is doing something else, even flying to the next flower, and so rapidly that the eye can't follow more than a fraction of one part of it during each observation.

Consider the concentration required when the bee master opens a hive of 60,000 bees. How strenuously he must hold his thoughts to the teeming world before him in order to perceive the order achieved. Here is a feat worthy of Holmes, requiring the remarkable concentration of which he alone is capable.

Workers progress through the jobs of the hive as they progress through their lives. With the expected apian efficiency, the worker bee's career moves from where she is born outward. She seems to follow a motto from Carlyle's *Sartor Resartus,* "Do the duty that lies nearest thee which thou knowest to be a duty!"

When the worker bee emerges from her brood stage, she starts her life by cleaning the hexagonal cell in which she was born. Then, after a day of orientation, she becomes nurse to newly-laid eggs, takes a turn at janitorial chores, cares for the queen, guards the hive, moving ever outward until she ends her life as a forager. The movement of workers from task to task is fluid and seems at least in part to be a response to the needs of the colony. Under very unusual circumstances worker bees can actually reverse the outward progress. In a depleted hive, for example, forager bees have been known to return to the hive to serve as nurses, forcing their atrophied glands to produce royal jelly once again.

Each queen reigns three or four, occasionally even five, years before the workers sense the need to raise a new queen. Since the only function of the drones is to fertilize the queen, and each queen requires that service only once, many generations of drones are born to no purpose. When the season of fecundation is concluded without the need for their services, the workers rid themselves of the useless drones. Either they are executed and removed, or they are forced from the hive where they will die within days. They will be prevented from joining another hive by the fact that they are marked, probably by pheromones, by their queen. Each hive is complete to itself, and bees from different hives can be caused to join together only under the most unusual circumstances. The exception is the honey-filled bee who comes from a hive that's been destroyed or moved. Bees are apparently as lenient to those who arrive with bribes of honey as was Cerberus.

We were all forced to learn as children exactly how the busy bee improves each shining hour: she is obsessed with work. This probably explains her eagerness to rid the hive of the loafing drones. Mr. Quinby stated,

> Industry is a part of bee nature. If, when their tenement is supplied with all things necessary to take them safely through the winter, and there is no necessity for continued labor, we furnish the additional room they assiduously toil to fill it up.

Indeed they don't seem to know what it means to have enough. As long as there is nectar in the flowers and a place in the hive to put it, they just keep bringing it in; this is what makes beekeeping so profitable. And it's not just the foragers who are devoted to labor: nurse bees are incredibly busy. Scientists have counted more than 1300 visits to a single egg on a single day! And the total of all visits to a single cell, egg to maturity, is approximately 10,000. 10,000 visits to raise a single worker bee–one of 60,000 in a healthy summer hive! But it should be noted that bees have idle hours as well. Observers have commented on foragers curling up in cells for a rest, and hive workers pacing up and down seeking a task.

This raises the fascinating question of communication among the bees. By the way the hive seems to act as one body, it is clear that communication is constant and effective. It doesn't seem so remarkable that workers rush to defend the hive if it is breached by an intruder, until we reflect that bees have no ears. Even more notable is the way foragers react to the needs of the hive, gathering the proportion of nectar and pollen appropriate to the specific requirements of her colony. How do they know? And the entire hive learns of the death of the queen within minutes. What means of communication are used?

Observers have not yet found all the answers, but it seems apparent that at least two varieties of communication take place in the hive: tactile and chemical. The first is well-known. Often called the bee dance, it is the method of passing on information on the location of good sources of nectar and pollen. The bee dance is so precise that once we learn to understand its code we can go directly (in a bee line) to the source of nectar every single time.

Dr. Karl von Frisch was the first man to understand the language of the bee dance. He is another in a marvelous gathering of remarkable people who have turned to the study of bees. His accomplishment should be seen in a line with Champollion's decipherment of the Rosetta Stone and Michael Ventris's discovery of a path through Linear B, an early form of Greek. Von Frisch could read the message of the bees every time, but he knew this accomplishment was so remarkable that the scientists of his time would not believe him. Indeed, he remarked, "No competent scientist **ought** to believe these things on first hearing." So he invited all inquiring skeptics to come to Munich to test his understanding of the bee dance. He had set up a massive testing ground with feeding stations at precise intervals radiating outward from the test hive. Dr. von Frisch stayed by the hive and invited the visiting scientist to pour out a measure of sugar syrup at the feeding station of his choice, using any subterfuge that would

make it certain he had not been seen. Dr. von Frisch would watch the first forager dance the location of the food, and then would tell the astonished observer precisely where the bees had been fed. He was always right.

The bee dance is performed inside the hive in near darkness. It involves an exact series of runs and wiggles and turns, and the vigor and enthusiasm with which it's performed carries its own message about the richness of the source being advertised. Other workers comprehend the location, primarily through their sensitive antenna, which bear organs of taste and smell as well as touch. As soon as they have the message, the workers pour out of the hive and fly directly to the food source. The bee dance uses the angle of the sun as referent, but even when it's hidden by clouds the bee is able to read the refraction of the light. Finding the desired location requires an abstruse computation in geometry, but that does not faze the tiny bee.

Bees are "flower constant." Over a period of several days, each worker will gather nectar from one kind of flower only. This constancy applies to the individual bee, however; others from the same hive may be foraging on a different flower. Honey labelled as derived from a particular blossom such as clover or thyme is really something of a mixture, though it will come primarily from the most widespread bloom of that place and time.

Bees have a keen and exact sense of place. When they fly out of their hives, they commit to their memories a precise picture of all the significant landmarks along their route. It is such a careful picture that if the hive is moved as little as ten feet away, it is effectively lost to the returning foragers. This same sense of place makes the precision of the directions imparted in the bee dance possible. The directions take into consideration wind velocity as well as distance and direction.

The chemical form of communication is more mysterious. All bees are marked with the pheromone of their hive, which originates with their queen. This marking is spread through the caretakers in their constant contacts with the queen as they minister to her needs, and from those workers outward to others until all of the 60,000 or so bees in the summer hive are marked. Communication, moving in the same manner, is phenomenally swift: it takes little more than an hour for the entire hive to know that their queen is gone or that the hive has been breached by a predator. And they manage this without telephone or radio—without so much as a megaphone—simply bee to bee to bee to bee, using their antennae and certain potent scents their bodies exude.

The queen produces two aromas: one is a powerful drone attractant produced by the virgin queen in her mandibular glands. It resembles milk and is extremely effective, summoning drones from a wide area. The second is her personal odor, known as queen substance, which permeates the hive and gives it individual identity. It is for exclusion, for hive protection. One element of this personal aroma may be the scent of the flowers that colony is currently working. Queen substance can be obtained from any part of the queen's body, and is passed to workers through her caretakers and thence through the natural activities of the hive. It sometimes appears to observers that worker bees actually become intoxicated by queen substance. When a queen becomes old or ill and failing, her production of queen substance becomes reduced. This signals her daughters that the time has come to supersede her. It is the desire for queen substance that holds the hive together.

Bees probably evolved in Africa, and they prefer tropical temperatures. But they are remarkably adaptable, and so they've learned to maintain a steady tropical warmth within their hives no matter what the outside temperature may be. In wintertime, they must generate heat and stay together in a tight cluster to keep the temperature the way they like it. Bees generate heat within the cluster simply by metabolizing fiercely. Honey is a dense, heat-producing carbohydrate food. One tablespoon of it contains sixty-five calories, and bees are efficient converters of those calories into heat. They are tidy little creatures and refuse to defecate within the hive. If the temperature rises above freezing they will fly out on what are euphemistically termed "cleansing flights"; if the weather remains too cold, they will sicken before they will foul their home.

Before winter strikes, bees empty honeycombs in the center of the hive, enabling them to form a nest surrounded by the vital honey. When the air grows cold they form a cluster to share warmth, and the cluster contracts automatically when the weather grows colder, exposing less surface to the surrounding chill. Each bee is drawn in toward the warm center of the cluster where the queen remains. This constant migration forces other bees outward, and thus the work and the warmth are shared and maintained, guaranteeing the survival of the colony.

Remarkable to recall is the fact that no bee in the cluster, with the possible exception of the queen, has ever experienced winter. How do they know how to proceed?

Bees are wonderful architects, constructing efficient chambers for their two vital purposes: storage of food and maintenance of brood. Each cell in the honeycomb is a perfect hexagon, the bottom having three sides. If one set out by mathematical means to discover the shape that a given quantity of matter must take to possess the optimum capacity and strength, occupying the least space and consuming the least labor in construction, the answer would be the hexagon. The circle is the most perfect shape for storage and for the roundish larva, but abutting circles share no walls and leave wasted space in the interstices. A hexagon is an angled circle, and each wall serves two cells with no unused space whatsoever. Mr. Hoyt asks,

> How does it come that the bees have solved so exactly this intricate problem, and know in just what form and shape their precious wax can be used so as to hold the most honey with the very least expenditure of labor and materials?

Dr. von Frisch agrees that the mystery remains, "How they arrived at this none of our learned men has so far been able to discover."

Instinct is inherited memory.

Charles Butler
The Feminin' Monarchi', 1609

Every Sherlockian should have deduced that Victorian bee masters were a remarkable gathering of uncommon persons. There is no need to seek evidence that Holmes met any of these individuals. It is enough to know he would have felt stimulated in their company, and undoubtedly was responsive to their writings. It will be my pleasure to introduce you to three remarkable gentlemen of the 19th Century: François Huber, Lorenzo Lorrain Langstroth, and John Allen, and to another who wrote his book as an old man in 1946. First, however, let me lead you on a brief excursion further back in time.

Beekeeping was first recorded in paleolithic times. Among a series of prehistoric rock paintings near Valencia, Spain, is one showing beekeeping, if that term can properly be applied to the plundering of a wild bees' nest. This work was created about 15,000 years ago. The fact that prehistoric man was prepared to brave the stings of outraged bees (for the painting shows the human harvester warding off a fierce attack) is a clear statement of the high regard he had for honey. And as the earliest civilizations emerged, honey took an important place in ceremony and in medicine. It served as remedy and sacrifice, sometimes simultaneously. Honeybees can be found preserved in amber of ungaugable antiquity, carved on the tombs of ancient Egypt, and used both literally and figuratively in the Old and New Testaments. I remain skeptical, but cruses of honey sealed in a Pharoah's tomb, and jars found in an ancient cave in Crete were said to be fresh and fragrant when uncovered thousands of years after they were made. Honey is a truly ecumenical substance since reference to its benefits can be found in pagan mythology, the Bible, the Koran, and the Talmud. The Koran contains a whole chapter entitled "The Bee," and according to the teachings of Mohammed the bee is the only creature ever spoken to directly by the Lord. In Hebrew, the honeybee is called Deborah, she that speaketh. Aristotle called honey "dew distilled from the stars and the rainbow," but he succeeded in describing the reproduction and development of the honeybee at a time when most people thought bees were generated from the

intestines of a dead ox. Pliny, Aristomachus, and Columella were classical bee experts of note, and the entire fourth Georgic of Vergil was devoted to beekeeping. Indeed, Greek and Roman literature abounds in references to bees and honey. When Pompey and a thousand men were marching through the mountains, their enemies placed crocks of poisoned honey along the route. Pompey's men ate it, became unconscious, and were attacked and killed. The bees themselves have been used as a defensive weapon. Amurat, Emperor of Turkey, having besieged Alba and made a breach in the walls, found the breach defended by several hives of angry bees. The Janissaries, the bravest militia of the Ottoman Empire, refused to advance into the breach.

Still, the most fascinating bee masters lived in the 19th Century as our Master did, and if we are to believe H. Malcolm Fraser that is no accident. He wrote:

> The opening of the nineteenth century may conveniently be regarded as a turning point in the history of beekeeping. Until 1800, science and practice in beekeeping had been considered completely distinct and unconnected.

One of the earliest bee men of science was Francois Huber, a citizen of Switzerland. While still in his teens he went completely blind. His fiancee, against the advice of family and friends, married him anyhow. She, along with an unusually patient and perspicacious servant named Burnens, became his eyes, for this indomitable gentleman did not allow his handicap to interfere with his chosen field of study. He brought to his experiments a rigorous scientific standard, and an ability for inventing and constructing ingenious trials, successfully isolating a single variable for study. So rigorous were his methods that he can be credited with the establishment of a large number of facts since the clarity and precision of his insight was seldom questioned. Although he wrote the words that follow more than a century before von Frisch deciphered the bee dance, Monsieur Huber foresaw its clear necessity:

> There is nothing repugnant to our conceiving the subsistence of a language among beings whose instinct is so enlarged as that of bees; whose conduct is compounded of a thousand proceedings, and which, combined in great numbers, cannot share

their respective labours, or aid each other suitably without the means of mutual communication.

His sightlessness points out, too, that the most important components of the observation of bees are done not with the eyes, but with the mind. Perhaps his blindness preserved him from distraction and error, for his findings ring true today, 170 years after his *New Observations on the Natural History of Bees* was first published in Geneva. He was clearly a gentleman and a scholar.

The greatest of the Victorian bee masters of America bore the memorable name of Lorenzo Lorrain Langstroth. He was not only a careful observer of bees, the discoverer of "bee space," and the inventor of the modern beehive, but he was also the author of *The Hive and the Honeybee,* first published in 1853, a book both instructive and graceful. With great effort, he imported and developed *Apis mellifera,* the Italian race of honeybees still preferred by most beekeepers in this hemisphere. He did all this in a single lifetime, yet by his own admission he was mentally unbalanced for half of his adult years. Mr. Langstroth was born on Christmas Day in 1810 in Pennsylvania, and took an extraordinary interest in observing the habits of insects starting at age six. He attended Yale where he was present at what may have been the first college food riot, the 1828 Bread and Butter Rebellion. In his *Reminiscences* he wrote that "the bread was not always sweet nor the butter fresh." He was later racked by guilt over his minimal role in the ruckus: he had promised his mother to be good. Mr. Langstroth was a Latin scholar who memorized a large portion of *The Aeneid.* Yale had no course in the curriculum relating to biology or nature. Therefore, for answers to his questions on bees, he turned to the hive itself—a most revolutionary approach. He went on to study for the ministry, supporting himself by teaching at several girls' schools in New Haven. When he attempted to preach his first sermon, he was seized by an hysterical voicelessness and found himself unable to deliver it. This was the start of what he later labelled his "head troubles." He turned to beekeeping as a healthful outdoor activity, but when he was suffering from one of his spells of melancholia, he could not bear to sit within sight of his hives, nor even to behold the letter B! He barely spoke, never raised his head, fasted, and shut himself away doing chess problems all the days until his depression broke. Some-

times it took months. Perhaps chess is the sign of a troubled mind, as well as a scheming one.

Reverend Langstroth's most basic discovery was the magic of "bee space," which is three-eighths of an inch. It was this discovery that made possible a beehive with movable frames and ushered in modern beekeeping. This is the ideal space to be left by the wise beekeeper between combs. If the space is narrower, the bees seal it with propolis, rendering the combs immobile without destruction. If the space is wider, the bees simply build more comb, again rendering much of the hive inaccessible without extensive damage. In his journal, Reverend Langstroth related how he came to an understanding of this phenomenon while walking home from his apiary. He admits, "I could scarcely refrain from shouting out my 'Eureka' in the open streets."

Reverend Langstroth wrote in a lively style that made scientific fact as fascinating as romantic fiction. His writings were enriched by his wide acquaintance with the classics, by his keen insight into the human heart, and by his unusually graphic imagination. His prose is delightful and deserves to be shared. In preparation for Reverend Langstroth's comments, it is necessary to describe the phenomenon of swarming. When the will of the hive dictates, the bees will swarm—a large proportion of the hive's population will suddenly fly out from their home to seek a new one. It happens all at once, though attentive beekeepers can predict it since the sound of the hive, the tone of the buzzing, changes a day or two before the event. Swarming occurs when the population is dense or when the queen senses a new queen being raised or because of some apian wanderlust—only bees know for sure. A majority of the bees gorge themselves with honey, enough for several days, and exit from the hive, congregating on a bush, a tree, the overhang of a building. To ill-informed humans they are a frightening sight, but in truth the well-fed bees are at their most docile. For a sample of his felicitous prose, read Langstroth's description of a swarm:

> Bees issue from their hives in the most peaceable mood imaginable; and unless abused allow themselves to be treated with great familiarity. The hiving of them might always be conducted without risk, if there were not occasionally some improvident or unfortunate ones, who, coming forth without the soothing supply, are filled instead with the bitterest hate against anyone

> daring to meddle with them. Such thriftless radicals are always to be dreaded, for they must vent their spleen on something, even though they perish in the act.

Reverend Langstroth was also fascinated by his observation that worker bees continue their duties until just hours before their deaths. He wrote:

> Place yourself before a hive and see the indefatigable energy of these industrious veterans, toiling along with their heavy burdens, side-by-side with their more youthful compeers, and then judge if, while qualified for useful labor, you ought ever to surrender yourself to slothful indulgence. Let the cheerful hum of their old age inspire you with better resolutions, and teach you how much nobler it is to die with harness on, in the active discharge of the duties of life.

The Reverend Langstroth died during a church service, just before he was to deliver a guest sermon. Clearly, he accepted his own good advice and followed the example of the honeybee. He was 85.

The name of John Allen's book, *The Blessed Bees,* was its first attraction, and the discovery that the copy in the collection of the New York Public Library was inscribed by Mr. Allen to Lorenzo Langstroth made it even more alluring. This account starts as the Mr. Allen is forced to leave college when his father dies. His father, a successful small farmer whose holding was some seventy miles north of Lansing, Michigan, had a few hives of bees. In his luggage the 17-year-old John had a handbook of bee culture he had bought as a gift for his father. As he joined his mother and younger brother and sister in working the farm, John read the practical handbook and became intrigued in the capacity for profit. He then proceeded to order every book and periodical in the field, spending (he kept careful accounts) $34.65, a large sum in the middle of the last century. He devoured everything, the articles repeatedly and even the ads with care. When he had studied every book and periodical available at that time, he made precise plans. He decided he must start with at least thirty hives to make a profit in his first year, and so he purchased them from farmers in the area, obtaining, as it happened, thirty-six. From a supplier he purchased at a much higher price one hive of

the golden Italian bees first imported by the Reverend Lorenzo Langstroth. With great care and attention and hundreds of hours of work, he substituted Italian drones for the native black bees in every hive, effectively converting his apiary to Italian in a single season. Neighbors laughed at the lad's long hours of labor with the foolish bees, but he laughed last as he ended his first season with 1700 pounds of surplus honey to sell. When the time came to transport the honey to the railroad depot, he took his own quiet revenge for the older men's scorn:

> I engaged nine of the neighbors to come with their large wagons, taking special pains to hire those who had the most to say of my youthful folly, and my new-fangled notions about bees.

John Allen's Linnswick Apiary showed a profit of 454% in its first year of operation when he was seventeen, and when he wrote his reminiscences, his apiary held more than a thousand hives.

For many of the years between the World Wars, C. N. Buzzard lived and kept bees in the south of France. Some of the visitors he chronicled were Rudyard Kipling, Lady Elizabeth Russell, H. G. Wells, and E. Phillips Oppenheim. His bee mentor was Professor Baldensperger, whom he called the grandfather of European bee experts. The Professor was said to live on honey and a little milk, and lived well into a vigorous and active old age.

Mr. Buzzard speaks of the freemasonry of beekeepers of all countries. Meetings were arranged, visits made, periodicals and letters were exchanged, and the friendly and uncompetitive relations bring to our minds the close-knit world of Holmesians. He notes that Hitler forbade German bee masters to attend the international meetings, perhaps to guarantee the Aryan bee would not be forced to mingle with lesser varieties. Or perhaps, Mr. Buzzard mused, Hitler felt the inherent challenge of a land flowing with milk and honey.

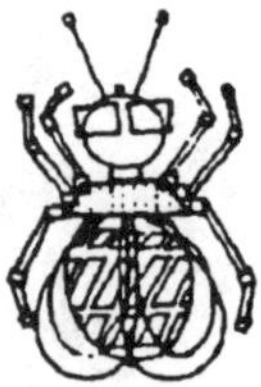

Like Sherlockians, beekeepers tend to cross-fertilize. They speak of each other with affection and respect, and it is clear from their journals that many visits took place and correspondence was frequent. I first met Francois Huber in the observations of the Reverend Langstroth, and weeks after I concluded the masterwork of Everett Phillips I encountered him again when he visited Buzzard. In *A Book of Bees,* Sue Hubbell shared the same sensation of being handed off from one bee master to another. She wrote:

> But once a person starts reading anywhere in the beekeeping literature, one author introduces the reader to another. One winter's reading generates the next winter's booklist.

Bees attract a most uncommon mind. That the science demands careful observation and rigorous recording comes as no surprise, but what is remarkable is the intellectual caliber of the Victorian bee masters. For the most part they were equally men of science and men of philosophy, towering minds bending their attention to the tiny honeybee. It's clear that beekeeping was highly regarded as an occupation. Here's Reverend Langstroth again:

> Beekeeping is regarded in Europe as an intellectual pursuit, and no-one who studies the wonderful habits of this useful insect will ever find the materials for new observations exhausted.

And here's a statement written by Reverend Langstroth about Monsieur Huber's work, but Sherlock Holmes comes immediately to mind:

> It would be difficult to find in any language a better specimen of the inductive system of reasoning, than Huber's work on bees, and it might be studied as a model of the only way of investigating nature so as to arrive at reliable results.

This is Langstroth on Huber, too, but one can almost smell the shag tobacco:

> His active yet reflective mind demanded constant employment; and he found in the study of the habits of the honey-bee, full scope for his powers.

In 1881 T. W. Cowan published *The British Bee-Keeper's Guide Book*, in which he observes:

> Although anyone may possess bees, it is not everyone who can become a proficient bee master. Only energy and perseverance, together with aptness for investigation, can ensure real success.

It seems Holmes found a worthy connection for his racing engine.

For so work the honeybees,
Creatures that by a rule in nature teach
The act of order to a peopled kingdom.

William Shakespeare
Henry V II, 3

Everett Franklin Phillips of Cornell calls the apian enthusiast a beecrank. True beecranks, like true Sherlockians, must possess a sense of history and a comfortable recognition of their place in it because there is so much continuity in both worlds. As we feel linked to the early organized Holmesians of the 30s and 40s, and past them to the time of our Master himself, the beecrank must feel linked to the great beemasters like Reverend Langstroth and Dr. Phillips, and Monsieur Jean Henri Fabre (whom Pope R. Hill, Jr. conjectured Holmes met during the hiatus in the south of France). There is continuity, too, in the equipment of beekeeping from the mundane to the complex. Pictures of the hive tool, the smoker, the queen excluder show very little change in the illustrated practical handbooks of the past 150 years. The

continuity of the instruments betokens the continuity of the practices.

One exception was the smoker of the German bee-masters around the time of Holmes's retirement. Beekeepers use smoke to disarm the bees when it is necessary to open the hive, and the Germans used a sort of two-headed pipe which could direct smoke into the hive at will. It is tempting to imagine Holmes combining his pleasure in tobacco with his passion for the observation of bees and utilizing such a tool, but beekeepers seem to agree that bees react better to "cool" smoke, and shag tobacco smokes hot.

Also contributing to the sensation of entering a timeless flow is the language of beekeeping, which is both graceful and apt. A beekeeper who has achieved wisdom and balance is known as a bee master, a term his followers should have no difficulty applying to Sherlock Holmes. The colony that is functioning and content and possessed of an energetic egg-layer has been called queenright for at least 200 years. The word rolls easily from the Twentieth Century tongue, but surely it was more comfortable still to an inhabitant of the Age of Victoria. Other more specific words have the right ring, too. Propolis is bee glue, produced by the workers from tree resin or even varnish.

Because it was first noticed in the structures the bees built in the wild to form the hive entrance, the substance was given its Greek name meaning "before the city." Supersedure is the rather chilling term given to the natural replacement of the queen, and virtually all the writers on bees characterize the workers as "drawing out" the cells of the honeycomb, a pleasingly graphic phrase. Many terms are a delight on the tongue or in the mind's eye: Spring dwindling, nectar flow, skyscraper hive, and foulbrood, a disease as dreadful as it sounds.

Beekeeping has provided a source of inspiration for great writers as well. Langstroth called beekeeping "the poetry of rural economy," and indeed *Apis mellifera* inspired one of Shakespeare's most poetic and musical lines, "the singing masons building roofs of gold."

The meaning of "bee" in the word beekeeping is becoming clear, but that seems to make the "keeping" portion decidedly puzzling. In what way can a person be said to **keep** bees? Certainly not in the financial sense, since all the handbooks are based on the unshaken premise that all over the world bees kept intelligently are kept profitably. Clearly, in a monetary sense bees keep their keepers. Then is there a relationship between keeping a bee and keeping a woman? Hardly, since the primary responsibility of a kept woman is to languish creatively and maintain a sort of eager availability, while the bee takes a very different attitude. Indeed, is "keep" the appropriate form for a human being's relationship to the honeybee? Practical and theoretical bee people agree that strenuous means are to be avoided, and that it's best to interfere in bee management as little as possible. So perhaps it is keeper in the sense of keep away.

As soon as an interested person begins to engage in research about *Apis mellifera,* the honey bee, he or she stands amazed at the enormous volume and variety of books, pamphlets, and periodicals that exist on the subject. After all, the first published works were written when languages like Classical Greek and Latin were alive and well. Thousands of volumes are available to even the most desultory researcher. So just when one is sure the well-observed honeybee has no privacy left, comes the realization that all the commentators are so vague about the nuptial flight because they are still uncertain of the details! There is disagreement on one of the most fundamental questions: how often does the queen mate during her single amorous excursion? Actual observation has been rare and fragmentary, even now when sophisticated equipment opens new possibilities. Only in the past decade or two have scientists become convinced that she mates many times, perhaps between ten and twenty. The queen bee may seem to us more mechanical than admirable, but it is impossible not to appreciate her achievement. For thousands of years she has kept what's private, private.

Though theories abound, scientists are not at all certain what causes bees to swarm. They do not really know how the division of labor is achieved—it seems both rigidly systematized and adaptable to the needs of the colony. The nature and derivation of the hive odor also remain a mystery, as does the function of some of the glands of the worker bees, and even some of the uses of the three simple eyes atop their heads. Humans have been looking at the honey-bee for 4000 years and there is still a lot we don't know. This astonishing complexity would pose a genuine challenge even to the capacious mind of Sherlock Holmes.

Since Holmes retired, countless apian experiments have been perpetrated and thousands of volumes have joined the library of bee culture. Understanding of

Apis mellifera has grown, but for those of us who relish hard questions, it is encouraging to see that many of the hardest still await solutions. Edwin Way Teale published *The Golden Throng* in 1940, but the intervening years have surely failed to invalidate his statement:

> Scientists who study the bee—chemists who analyze the content of royal jelly, microscopists who attempt to count the sperm cells in the pouch of the fertilized queen, physicists who measure the stresses of the honeycomb, all reveal the physical characteristics of the bee and its surroundings. But beyond this lies that intangible, mystical force, the spirit of the hive, running through all the activity of the golden throng; ordering the lives of the insects, molding them into the compact, unified, efficient commonwealth we know.

In our eagerness to explore the intellectual content of bee culture, we must not ignore the most practical of all the bee's offerings: pollination. Many of our staple crops would not exist without the bee as go-between: clover, alfalfa, and all our orchards, to cite a few. And, as one gift sometimes gives twice, this need for the bees' notice has given us the glorious colors and aromas of the flowers. It is reasonable to assume that the colored petals and alluring scents of flowers are an advertisement, showing bees and other winged creatures where something good is to be had, and where their visit is required.

The golden bounty of the honeybee deserves notice as well. Honey is both an antiseptic and a beauty aid, but most important, it is a sweet and glorious food. It is absorbed into the blood-stream faster and easier than any other sweet. It does not need to go through the full human digestive process because the sugars start to break down in the bee's honey-stomach and require little further processing. Honey can properly be called predigested, but the squeamish may prefer to forget by whom.

Cecil Tonsley reminds us that,

> Honey is, however, a complex substance, varying appreciably in its composition from flora to flora, country to country, soil to soil, and is even affected by climatic conditions.

Indeed, honeys are distinct, and those who taste them with educated care have many of the same criteria as oenophiles: color, aroma, vintage, and savor. They sound rather like wine connoisseurs as well, though without the elevated vocabulary. For them, no rose is an extra—each floral varietal commits some of the nectar that the bees will transform into golden honey.

And live alone in the bee-loud glade.

William Butler Yeats
"The Lake Isle of Innisfree"

Some purely practical reasons for Holmes to undertake the keeping of bees are provided by the fourth point raised in the preface to *His Last Bow*. It notes, you remember, that he was "somewhat crippled by rheumatism." Here was a man ready to change his life, prepared to forswear the comforts and amusements of his London life to devote himself solely to the pursuit of the mysteries of nature, when he found himself suffering from certain pains. He examined his symptoms with his clear and rational mind and made a diagnosis. He understood the implications of his discomfort immediately—after all, he studied in a teaching hospital, though his studies in medicine were far from systematic. He would keep his medical knowledge current, too, because an ability to recognize the effects of all known diseases was important in his first career. So he knew he was right when he recognized in his own joints the earliest signs of rheumatoid arthritis, an inflammation that could stiffen his limbs and deny him the flexibility imperative to a naturalist. What could he do? His mind was prepared to take on the greatest of natural mysteries, but he must face the possibility that his body was likely to restrict the quest. How could he manage it?

It is my contention that he found the answer in a small book called *Bees, their Habits, Management, and Treatment.* Its author, a well-known naturalist, died in 1889, and we know the Master was familiar with the author, the Reverend J. G. Wood, since he owns and knows well another book by the same man, *Out of Doors*. That was the book he uncovered in his attic in "The Adventure of the Lion's Mane," that revealed the deadly killer, *Cyanea capillata.* In Reverend Wood's book on bees we find a great deal of good practical advice, including the following words:

> He who desires to see the lion, the elephant, or the giraffe in the full enjoyments of their natural tastes, engaged in their natural pursuits and following

their natural instincts, must . . . leave his native land and spend years separated from his fellow men, in the depths of the African forest, or on the wastes of a burning desert . . . Animals to be studied properly must be studied in their natural states, and if the observer cannot go to them, he must bring them to him. This he cannot do with the larger animals, but he can with the insects. The space which would not suffice for a lion to turn in, will form a forest for a beetle . . . a world for a bee.

And so he brought all of nature to him and built a bee yard. He brought all that mystery within an easy stroll of his cottage's door, so that as his rheumatoid arthritis progressed it would remain accessible to him. He brought great questions to the limits of his own backyard: the amazing communication among bees, the incredible efficiency of the organization of the hive, the constant maintenance of balance, the miracle of the making of a queen, and the vaulting nuptial flight. He brought all this to him, yes, even a taste of honey. He built for himself a complete laboratory of nature, and in so doing uncovered his own cure.

Nearly all the books about bees mention it. It doesn't seem to matter if their thrust is practical, scientific, or philosophical; it doesn't matter if they were written in this century or an earlier one. Most stress that the theory remains unproven, but all have seen examples of its efficacy. It was mentioned in Reverend Langstroth's volume:

> The poison of the bee has a beneficial effect as a cure for rheumatism. Numberless cases have been reported where a few stings have caused the disappearance of this affliction.

In the *Biggle Bee Book,* first published in 1913, Jacob Biggle seems to entertain few doubts about this or other matters:

> . . . the poison of the bee sting is an invaluable remedy for rheumatism . . . It has been known for many centuries that, generally speaking, beekeepers are immune from this affliction, due, no doubt to the agency of the bee sting.

But Colin Butler, writing in the middle of this century, is more circumspect when he states,

> There is no doubt that treatment with bee venom sometimes achieves spectacular results in the alleviation of rheumatism, but alas, in many cases it fails in its object.

Published only a few years earlier, *Bee's Ways* by George DeClyver Curtis gives specific directions on the subject:

> One other curative product of the bee, not so pleasant as honey but highly effective, is the venom of the sting. For many years it has been a well-known fact among beekeepers that none of them ever suffers from arthritis or rheumatism, and it is equally well known that they have cured members of their own families and neighbors by putting bees on the swollen joints and provoking them to sting.

Bringing us to the present, Sue Hubbell describes the same technique in her recent *A Book of Bees*:

> Like many beekeepers, I have discovered a dose of bee venom from a sting alleviates the symptoms of arthritis, and I am stung so infrequently that when the joints in my hands begin to ache I have to go capture a bee and force her to sting the place where the hurt is. The pain from the arthritis is gone by the next morning.

It is clear that Holmes made the same discovery, and the rather unpleasant technique would have been no difficulty for a man so well-versed in the use of hypodermic needles.

Compound of the Busy Bee and Excelsior. We can but try—the motto of the firm.

Sherlock Holmes quoted by John H. Watson
"The Adventure of the Creeping Man"

Excelsior and the busy bee: it's exactly what we want for Sherlock Holmes. Excelsior asks that he strive for his best, and the bee provides a worthy focus for intense study. Once again Holmes demonstrates personally how to handle a time of change. He inspires us anew with the faith that we can take charge of our lives, can shape the change that's become necessary. There are some things humankind can never have enough of, and surely a true example of grace under pressure is one of them. In 1891 Holmes, facing death, expressed his desire to turn his thoughts to nature. His mind darted around the world and through the sciences seeking a road to follow. Even as his thoughts ranged outward, the voice of his body forced him to look within. His pain and growing stiffness caused him to seek a study close to home, and so he chose the fascinating and enigmatic bee. This was precisely the right choice for

Sherlock Holmes. He could continue to lie in the center of millions with his filaments stretching out and running through them. He could still feel every vibration, but now the vibrations would be accompanied by buzzing instead of imprecations. There would be no stagnation to rebel at, no idleness to irritate him, and wondrous mysteries awaiting solutions.

It is certain that the subject of his study was entirely worthy of this best and wisest man. According to Princeton biologist James L. Gould and his partner, Carol Grant Gould, the honeybee is, in its context, as advanced an organism as a human. A billion years ago, animals divided into two groups. One group developed seventeen phyla, including the arthropods of which insects are the dominant class. The other evolved into only four phyla, among them vertebrates, to which we belong. "Honeybees are at the top of their part of the evolutionary tree," wrote the Goulds, "whereas humans are the most highly evolved species on our branch. To look at honeybees, then, is to see one of the two most elegant solutions to the challenges of life on our planet."

APICULTURAL BIBLIOGRAPHY

With personal notations

Allen, John *The Blessed Bees* New York, G. P. Putnam's Sons, 1878
Charming, personal account of a young Michiganian

Biggle, Jacob *The Biggle Bee Book,* A Swarm of Facts on Practical Beekeeping, Carefully Hived, Philadelphia, Wilmer Atkinson, 1913
Practical and folksy, if occasionally heavy-handed

Bonsels, Waldemar *Maya: the Adventures of a Bee* (translated by Charlotte Remfry Kidd) London, Hutchinson's Books for Young Persons, 1922
A children's book with a thoroughly silly text, but charming illustrations by Homer Ross

Butler, Colin G., M.A., Ph.D. etc. *The World of the Honeybee* New York, The MacMillan Company, 1955
Clear, illuminating, considered

Buzzard, Charles Norman, *Shining Hours* London, Collins, 1946
Keeping bees as WWII crept into southern France —written by a wise and witty gentleman with humor and fondness

Crompton, John *A Hive of Bees* New York, Nick Lyons Books, 1957
An intelligent, personal perspective with real charm

Curtis, George deClyver *Bees' Ways* Boston, Houghton Mifflin Co., 1948
Readable, meandering—the author is pleasant company.

Fabre, Jean Henri *The Mason Bees* (translated by Alexander Teixeira de Mattos) New York, Garden City Publications, 1925
Masons are an interesting contrast to the more sophisticated *Apis mellifera*

Fraser, H. Malcolm *History of Beekeeping in Britain* London, Bee Research Association, Ltd., 1958
Mildly interesting perspective

Grout, Roy A., editor *The Hive and the Honeybee* Hamilton, Illinois, Dadant & Sons, 1949
One of several updates of Langstroth—unless you're on the verge of purchasing hives, read the original.

Hay, Vicky "Arizona's Sweet Bees" *Arizona Highways*, March, 1990
Concise interesting narrative with wonderful photographs—includes interview with the Goulds of Princeton.

Hoyt, Murray *A World of Bees* New York, Bonanza Books, 1965
Folksy, sometimes awkward, but always enthusiastic—Hoyt's sense of humor keeks through.

Hubbell, Sue *A Country Year* New York, Random House, 1983

——*A Book of Bees* New York, Random House, 1988
If you pick up either of Sue Hubbell's books you'll thank me forever. I'm sure of it.

Huber, Francois *New Observations on the Natural History of Bees* Edinburgh, W. C. Tait, 1821 (3rd edition)
Clear, cogent, sometimes inspired—doesn't seem to be 170 years old

Langstroth, Lorenzo Lorrain *A Practical Treatise of the Hive and the Honey-Bee* Philadelphia, J.B. Lippincott & Co. 1868 (3rd edition)
A completely rewarding book to read

Larson, Peggy Pickering Mervyn W. *Lives of Social Insects* Cleveland, Ohio, The World Publishing Company, 1968
Well organized and clearly written—good to read of the social organization of other insects as well

Longgood, William *The Queen Must Die,* and Other Affairs of Bees and Men, New York, W.W. Norton & Co., 1985
Sometimes Mr. Longgood's insistent anthropomorphizing is irritating, but bees provide a real temptation in that direction. His voice is individual and the book is good company.

Maeterlinck, Maurice *The Life of the Bee* (translated by Alfred Sutro) New York, Blue Ribbon Books, 1901
Desperately overwritten, but still charming

More, Daphne *Discovering Beekeeping* Haverfordwest, England, Shire Publishing, Ltd., 1977
A pamphlet: straightforward, pragmatic

Morse, Roger A. *Bees and Beekeeping* Ithaca and London: Cornell University Press, 1975
Well structured, clear, readable, and authoritative

Naile, Florence *The Life of Langstroth* Ithaca, Cornell University Press, 1942
Dumb: solicitous when he did not require solicitude

Ohio State Archaeological Historical Quarterly "Langstroth, the 'Bee-Man' of Oxford" Columbus, 1948
Derived primarily from Naile

Phillips, Everett Franklin *Beekeeping, A Discussion of the Life of the Honeybee and of the Production of Honey* New York, The MacMillan Company, 1915
Weighty, practical, detailed, opinionated

Quinby, Moses *Mysteries of Bee-Keeping Explained* New York, Orange Judd & Co., 1866
Mr. Quinby is a testy fellow. Rather a period piece

Riches, H.R.C., M.D., F.R.C.P. *Beekeeping* London, Foyles Handbooks, 1976
As brief and practical as its terse title

Root, A.I. *Gleanings in Bee Culture* Medina, Ohio XIX #18, September 15, 1891
A pleasant dip into day-by-day Victorian bee culture

——*The ABC of Bee Culture,* A Cyclopaedia of Everything Pertaining to the Care of the Honey-bee; Bees, Honey, Hives, Implements, Honey-plants, etc. Facts gleaned from the Experience of Thousands of Bee-Keepers All Over Our Land. Afterward Verified by Practical Work in Our Own Apiary, Medina, Ohio The A.I. Root Company, 1905
What can I add?

Stratton-Porter, Gene *The Keeper of the Bees* Garden City, Doubleday, Page, & Company, 1925
A very old-fashioned and contrived novel with a core of genuine sweetness appropriate to the subject

Taylor, Richard *The Joys of Beekeeping* New York, St. Martin's Press 1974
Another first-person narrative from a gentleman of wit with an emphasis on the spiritual aspects of beekeeping

Teale, Edwin Way *The Golden Throng* New York, Dodd Mead & Co., 1940
A delight—a fine nature writer meets a subject that clearly pleases him.

Tonsley, Cecil *Honey for Health* New York, Charter Books, 1969
Occasionally silly, but fascinating—the work of a persuasive honey-crank

von Frisch, Karl *Man and the Living World* (translated by E.B. Lowenstein) New York, Harcourt, Brace & World, 1949

Probably written for young adults—this scientist has a real gift for making difficult concepts clear.

—— *The Dancing Bees,* An Account of the Life and Senses of the Honey-Bee (translated by Dora Ilse) New York, Harcourt, Brace, 1953

This is a luminous and exciting book to read.

Wood, J.G. *Bees, Their Habits, Management, and Treatment* London, George Routledge & Sons, 1853

Every Holmesian should read it, but it's not easy to find.

A Clear Concise Refutation of Other Commentators' Opinions About the Retirement of Mr. Holmes *and other stories*

Of the sixteen commentators I have rooted out who have chosen to focus on Mr. Holmes in retirement, virtually all of them posit some other occupation than beekeeping. Many of them mention the presence of hives, but for most beekeeping is either a blind for secret activities or an incidental hobby. Perhaps because they felt it impossible that the honeybee could draw an intellect of the calibre of Sherlock Holmes's, they neglected to explore apiary studies. Sadly, they were not stung with the fascination of the bee and thus failed to follow the Master.

Perhaps the earliest conjectures to appear in print can be found in P. M. Stone's "Sussex Interview," which appeared in *221B: Studies in Sherlock Holmes,* published in 1940. Like Starrett's "The Unique Hamlet," reprinted elsewhere in the volume, it is a pastiche. Clearly, the same weight cannot be granted to it as to recorded fact or even reasoned supposition since it is a loving work of fiction. It does mention that the Master possesses some conical beehives, but he's actually engaged in writing *The Whole Art of Detection* and continuing his chemical researches. Few questions are raised or answered in this comfortable piece of conjecture.

It is fortunate that scholarly matters are not settled by a vote, for the theory that has drawn the majority of the Sherlockian writers claims that after 1903 Holmes engaged in some form of espionage. This claim is based, of course, on the statement in the Preface to *His Last Bow* that "The approach of the German war caused him, however, to lay his remarkable combination of intellectual and practical activity at the disposal of the government with historical results which are recounted in *His Last Bow.*" While the Preface gives no hint that Holmes engaged in espionage other than that recounted in this tale, the statement provides a jumping-off place for further conjecture. Professor John B. Wolf felt that beekeeping was merely a blind for darker activities, and reminded us that Holmes lived in a villa near the sea, ideal for covert communication. T. S. Blakeney was certain that the Master was a key figure in the British Secret Service. Arthur L. Levine and Professor Frank Condie Baxter also believed that the only

pursuit of sufficient importance to draw Holmes away from Baker Street and detection was espionage against the hated Hun. Mrs. Creighton Sellars actually claimed that Holmes landed secretly with the troops at Normandy, never mind that he was ninety and suffering with rheumatoid arthritis.

More recently two eminent Sherlockians have joined the ranks of those who place Holmes on the dark paths of espionage, but that's virtually the only opinion David Hammer and Jon L. Lellenberg hold in common. Each has Holmes active as a double agent, but Hammer's Holmes ends up in Germany, Lellenberg's in Connecticut. Neither portrays Holmes as beekeeper.

Easily the most astonishing theory thrust forward by the espionage crowd was Norman V. Ballou's, "Top Secret," published in *The Third Cab.* He claims that from about 1920 to the beginning of the Second World War, Holmes lived continuously in Northern India. Here he succeeded in discovering, at least partially, the secrets of life and death, and learned how to extend indefinitely man's alloted span. This last was accomplished by preparing an elixir from the crushed bodies of young queens of the Oriental Hive Bee (*Apis indica*), the largest of the honeybees, common throughout India, Ceylon, and Malaysia. But that's not all. Holmes also found time to dominate the planning and espionage activities for two great governments, to continue his chemical researches, and to consult Buddhist scholars in Tibet on his study of the principle of life. When World War II broke out he went to live with the Dalai Lama in Lhasa; perhaps then he was permitted to rest.

It was William S. Baring-Gould, however, in his biography of the Master who formulated the widely-quoted theory that Holmes found the secret of his extreme longevity in royal jelly, the remarkable substance extruded by the worker bees. In the Baring-Gould biography, Holmes notes that on a diet of royal jelly alone, the larva of the queen bee increases in weight 1500 times in just five days. And this is the substance that has the power to change a common worker into an egg-laying queen. Baring-Gould does not deign to describe the master's actual chemical experiments with royal jelly, but he promises that Holmes has sent his conclusions to the Royal Society. We continue to wonder why this miracle has not been vouchsafed to the world at large.

There was also support in the Higher Criticism for the continuation of other chemical research during the Master's retirement. P.M. Stone's pastiche gives us a Holmes still happily pursuing truth at his deal-topped table. Mr. R. P. Graham felt his need for more time for chemistry was the reason he left his life as a consulting detective, and J.H. and Humfrey Michell posit a life spent among coal tar derivatives.

In his essay on the retirement years published in 1974 in *The Noble Bachelors' Red Covered Volume,* Bartlett D. Simms proposes a unique theory. He believes that Holmes became a fisherman when he retired, a serious fisherman. Through this contact with nature, he came to realize the danger to the ecosystem posed by the swift escalation of industry, and devoted himself to the study of what we now call ecology.

Only three Sherlockian commentators appear to accept Watson's repeated statements that Holmes became a beekeeper. The earliest of the trio is Pope R. Hill, Sr. who published his article in the *BSJ* in 1947. He seems to feel that Holmes sought answers to important philosophical questions through the observation of bees, though he does not really pursue that thesis. He is more intent on demonstrating what turned Holmes to bees, and believes that moment came during the

Great Hiatus. While Holmes was in Montpellier disporting himself with coal tar, the great bee master Jean Henri Fabre was living just seventy-five miles away. Mr. Hill believes they met, perhaps by chance, perhaps because Holmes knew of his work and sought him out. It should be noted, however, that Fabre concentrated on mason bees and bramble bees and had little to do with honeybees.

Remsen T. Schenck published his memorably named "Holmes in Hymettus" in the *BSJ* in 1959. While Mr. Schenck does assume that Holmes's new career was indeed beekeeping, he theorizes that the bees are merely the experimental means to a study of genetics. His case is weakened by the erroneous assumptions about bees on which he bases his theory. His errors are so fundamental as to undermine his entire thesis. He does not appear to know that the queen can mate with whatever drones she encounters, not just those from her own hive. Nor does he seem aware that all bee brood is fed royal jelly, but only the developing queen is fed it exclusively and in great quantities. And his application of genetics to the study and practice of beekeeping seems rather, well, elementary. Mr. Schenck was also responsible for the limited production of a small blue book with its title inscribed in gold, *A Practical Handbook of Bee-Culture With Some Observations Upon the Segregation of the Queen.* The title page names the author as Sherlock Holmes and it is dedicated to John H. Watson. It is a guide for beekeepers.

A year later, W. H. Chenhall's piece on the retirement of Sherlock Holmes was published in the *Sherlock Holmes Journal.* Much of the paper is devoted to locating his villa near Eastbourne and refuting Christopher Morley's attempt to do the same. Mr. Chenhall clearly believed that the Master's retirement was a genuine one, necessitated by the crumbling of his intellectual powers. He was also disoriented by the departure of Watson who, Mr. Chenhall theorized, married a woman who did not approve of peremptory summonses from Holmes. He did become a beekeeper, not to observe bees and to seek truth, but because he was no longer capable of anything more complex. Mr. Chenhall believed that Holmes's rapid disintegration was caused by the depredations of cocaine and tobacco.

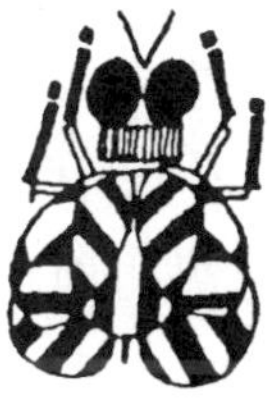

I submit that none of these commentators, though they were all men of learning, troubled to explore the world of *Apis mellifera* through book or through hive.

Even the most elementary acquaintance with the honeybee makes Holmes's fascination completely understandable, and renders unnecessary the need to find a hidden meaning behind the clarity of Watson's words. I join with Moses Quinby in asking,

> Is there in the whole circle of nature's exhaustless storehouse, any one science more inviting, more refining, and more exalting than this?

HOLMESIAN BIBLIOGRAPHY

Related Items from the Higher Criticism

Ballou, Norman V. Top Secret *The Third Cab* Boston: The Speckled Band 1960

Baring-Gould, William S. *Sherlock Holmes of Baker Street*: A Life of the World's First Consulting Detective New York: Clarkson N. Potter, Inc. 1962

Baxter, Frank Condie Foreword to *The Hound of the Baskervilles* Garden City, New York: Doubleday Page and Company, 1926

Blakeney, T. S. *Sherlock Holmes: Fact or Fiction* London: John Murray, 1932

Chenhall, W. H. The Retirement of Sherlock Holmes *The Sherlock Holmes Journal,* Vol. 5, No. 1, Winter, 1960

Graham, R. P. Sherlock Holmes in Retirement *The Baker Street Journal,* Vol. 1, No. 4, October, 1946

Hammer, David L. Sherlock Holmes, Secret Agent *The Baker Street Journal,* Vol. 36, No. 4, December, 1986

------------ The Twenty-Second Man *The Baker Street Journal,* Vol. 38, No. 1, March, 1988

Hill, Pope R., Sr. Sherlock Holmes Meets Jean Henri Fabre *The Baker Street Journal,* Vol. 2, No. 1, January 1947

Lellenberg, Jon L. The *Magnum Opus* of His Later Years *The Baker Street Journal,* Vol. 37, No. 2, June, 1987

Levine, Arthur L. A Man of Formidable Resourcefulness: An Essay in the Manner of a Footnote to History *The Baker Street Journal,* Vol. 4, No. 3, July, 1954

Michell, J. H. and Humfrey Sherlock Holmes, the Chemist *The Baker Street Journal,* Vol. 1, No. 3, July, 1946

Schenck, Remsen T. Holmes in Hymettus *The Baker Street Journal,* Vol. 10, No. 3, July, 1960

Sellars, Creighton A Visit to Sherlock Holmes *The Baker Street Journal,* Vol. 2, No. 1, January, 1952

Simms, Bartlett D. Sherlock Holmes, Proto-Ecologist *The Noble Bachelors' Red Covered Volume* St. Louis: Birchmoor, 1974

Stone, P. M. Sussex Interview *221B: Studies in Sherlock Holmes* New York: The Macmillan Company, 1940

Wolf, John B. Another Incubus in the Saddle *Exploring Sherlock Holmes* Ed. E. W. McDiarmid and Theodore C. Blegen LaCrosse, Wisconsin: Sumac Press, 1957

AFTERWORD

More than one revered teacher has expressed impatience with the kind of pupil who is quick to express admiration but reluctant to follow a regimen. In the foregoing monograph, Susan Rice has shown (not for the first time) that she is a true disciple of Mr. Sherlock Holmes: she knows his methods and has applied them.

When studying a body of evidence, we have been taught that, if one fact appears to be at variance with the rest, its correct interpretation may change our understanding of everything. While it may be human nature to ignore a puzzling fact or an uncomfortable one, Holmes has said, "The very point which appears to complicate a case is, when duly considered and scientifically handled, the one which is most likely to elucidate it."

Holmes's retirement has long been a stumbling-block for many of us. As the opening chapter here reminds us, our bewilderment at Holmes's dismissal of his sleuthing career and retirement to a life of agriculture has disoriented us. We don't know what to make of Holmes's decision, so we have made nothing of it.

The truth is that any thought of change to our image of Holmes disgruntles us. But what is this image? With a happy disregard of historical accuracy, we have pieced together a supposed norm of Holmes's existence—a composite whose elements were never all simultaneously present. We would have Holmes mature but energetic, with a constant flow of cases but plentiful spare time for esoteric studies, his fame broadcast throughout the Empire yet his ability to roam London anonymously unhampered; with Watson resident at 221B but neither too young nor yet bereaved; with Moriarty vaguely in the offing; with Baker Street Irregulars and cabinet ministers jostling on the stairs; with far more hansoms than Fords, and far more commissionaires than telephones.

But this isn't realistic: the years flowed by in Baker Street as they do for us. Society evolved and technology advanced and individuals aged. For us to desire a static Holmes is to treat him as if he were a fictional character, capable of acting out only one story in only one way. Susan Rice has done Mr. Holmes the honor of treating him like a real person, and he has reciprocated by letting her into his thoughts and his work.

Through her research, we come to understand how Holmes has made available to us the principles of perception and logical action that have guided him through his long life. Just as the youthful Holmes had created his unique profession of consulting detective to make the fullest use of his mental and physical abilities, and the Canonical Holmes constantly augmented his skills and knowledge, the Holmes of later middle age assessed anew both himself and his world, to find a new and worthy pursuit for the years to come.

The Canon's statements about Holmes's retirement form not a stumbling-block after all, but a guidepost. "Education never ends, Watson," he observed, and acted on his belief. With this insight, we can accept with good cheer the fact that it is, in fact, *not* always 1895.

—Evelyn A. Herzog, ASH, BSI
Principal Unprincipled Adventuress

Susan Rice

A.S.H.: Practical Handbook of Bee Culture with Some Observations Upon the Segregation of the Queen

B.S.I.: Beeswing

An enthusiastic Holmesian, Susan Rice is a member of (in alphabetical order) the Adventuresses of Sherlock Holmes and the Baker Street Irregulars, as well as a clutch of scion societies in several states. She received the Gaslight Award in 1983 for "Woman's Place in the Holmes," published in the *Baker Street Miscellanea,* and in 1990 shared the first John Bennett Shaw Award with good friend Evelyn Herzog. Susan has succeeded Lisa McGaw as host of the William Gillette Memorial Luncheon.

Susan is a fifth-generation Detroiter who has also lived in Ohio, Greece, Chicago, and now Manhattan. She taught at the Cranbrook Schools in Michigan, owned a shop called "Socrates" in Athens, and currently works in marketing for Celebrity/Fantasy Cruises.

She is an orthodox Holmesian.

Thomas Gentille

Back Door Sherlockian

Thomas Gentille is an unorthodox artist whose work takes many forms. He has had major exhibitions at the Asheville Museum in North Carolina, the Gulbenkian Museum in Lisbon, and the Victoria and Albert in London. He is currently working on a dozen projects in as many media, as is his wont, will, and way.

Other titles available from Gasogene Press

0-938501-01-1	*Immortal Sleuth*—Harrison	$24.95
0-938501-02-X	*Good Old Index*—Goodrich	$27.95
0-938501-00-3	*For the Sake of the Game*—Hammer	$23.95
0-938501-03-8	*The Game Is Afoot*—Hammer	$15.95
0-938501-04-6	*London by Gaslight*—Harrison	$24.95
0-938501-05-4	*The Ragged Shaw*—Shaw	$12.95
0-938501-07-0	*Studies in Scarlet*—Monographs	$19.95
0-938501-08-9	*The Twenty-Second Man*—Hammer	$15.95
0-938501-09-7	*Ms Holmes of Baker Street*—Bradley/Sargent	$19.95
0-938501-11-9	*To Play the Game*—Hammer (cloth)	$39.95
0-938501-12-7	*To Play the Game*—Hammer (paper)	$19.95
0-938501-13-5	*Doggerel in the Night-Time*—Sage	$10.95

All titles are available by writing

Gasogene Press, Ltd.

Box 1041
Dubuque, IA 52001-1041
319/583-3730

Please enclose $2.50 to cover shipping and handling for each title ordered to a maximum of $5.00. Iowa residents add 5% sales tax.